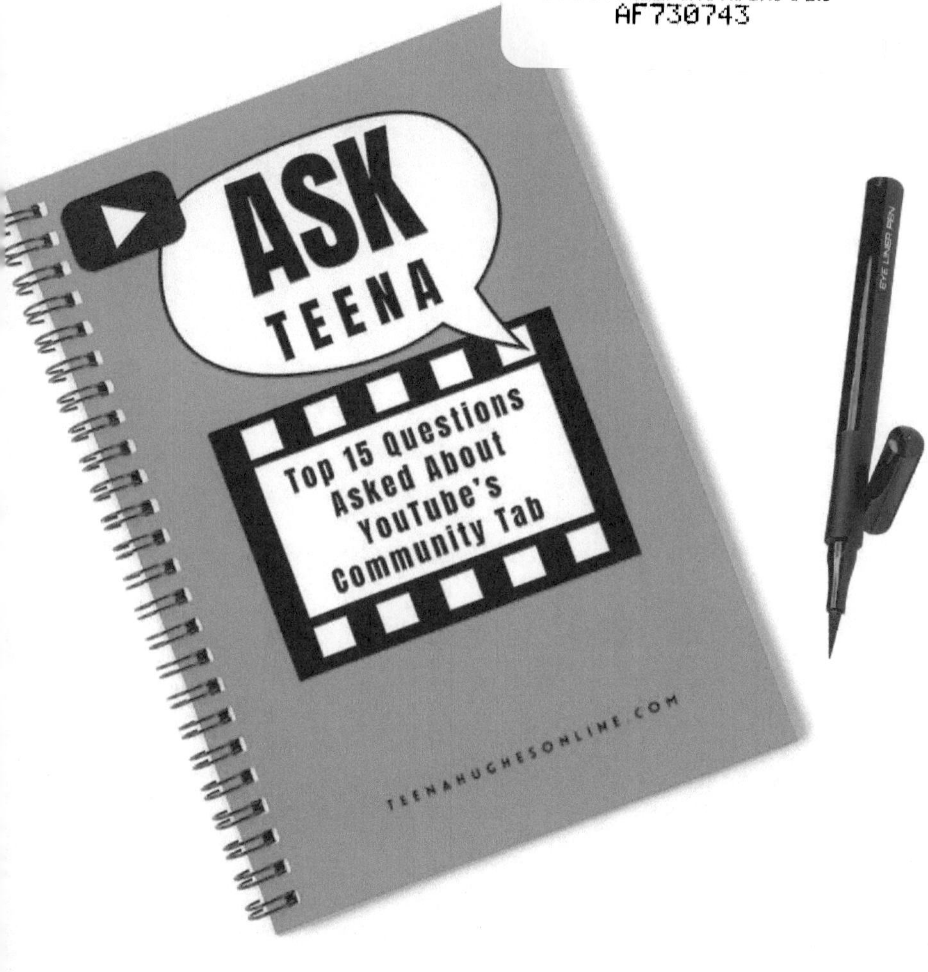

Questions, Answers and Notes Pages

This Book Belongs To

Website & YouTube Channel

Title: Top 15 Questions Asked About YouTube's Community Tab
"Ask Teena" Series -- 1st edition

ISBN: 978-0-9943972-4-9

Author: Teena Hughes, TeenaHughesOnline.com

Publisher: 42MonkeysPublishing.com

TABLE OF CONTENTS

Notes

Introduction

Hello! My name's Teena and I first discovered the Community Tab on my YouTube Channel quite by accident. I visited a few other Channels I follow, and clicked the link to see how they were using this new feature.

I was amazed and in awe! How clever is this? A new way to chat and connect with viewers and followers, where I can post photos, images, quizzes, GIFs and much more - I couldn't believe my luck :D

As I told more people about the Community Tab, I started to receive questions - lots of them - and the idea for this book was born.

It's not a training manual, it's **a straightforward way to answer** the top 15 questions I've received, with a view to clarifying the mystery.

- I've included **Notes pages** throughout, and I'm sure you may want to write down TONS OF IDEAS, and maybe even a few questions :-)

Please do enjoy this Booklet, and keep an eye out for more in the "ASK TEENA" Series.

I look forward to hearing from you if you try any of these suggestions on your YouTube Channel.

Wishing you the very best, Teena!

Introduction

QUESTION 1

What is the YouTube Community Tab?

The YouTube Community Tab is a feature that allows creators to engage with their audience directly on the YouTube Channel page.

It serves as a hub where you can post:

- updates (Posts)
- polls
- images
- GIFs, and more to foster interaction and **build a community** around your content.

This tab appears alongside your video sections, playlists, and channel information, making it easily accessible to viewers.

The Community Tab enables you to:

- share behind-the-scenes glimpses,
- tease upcoming videos,
- gather feedback,
- and cultivate a stronger connection with your loyal subscribers.

Notes

QUESTION 2

How easy is it to set up the Community Tab?

Setting up the Community Tab is a straightforward process.

Once you've enabled it in your YouTube Channel Settings, you'll find it prominently displayed on your Channel page.

The interface is intuitive and user-friendly, allowing you to create posts by selecting the desired format (text, poll, image, GIF) and adding your content.

- text
- polls
- images
- GIFs.

You can even schedule posts in advance or access analytics to gauge the performance of your Community Posts.

With a few simple clicks, you can start leveraging this powerful engagement tool.

This video is a quick overview:

- https://www.youtube.com/watch?v=wF_lsScEArw

Notes

QUESTION 3 Why would I bother using the Community Tab? What's in it for me?

The Community Tab offers numerous benefits for Video Creators looking to strengthen their connection with their audience and grow their YouTube Channel.

By regularly sharing engaging content and interacting with your viewers, you can **build a loyal community of dedicated fans who feel invested in your Channel's success**.

This increased engagement can lead to:

- higher watch times
- better retention rates,
- and ultimately, more subscribers.

Additionally, the Community Tab allows you to gather valuable feedback and insights directly from your audience, helping you tailor your content to their interests and preferences.

It's a powerful tool for fostering a sense of community, boosting discoverability, and driving long-term growth for your YouTube Channel.

Notes

QUESTION 4

Which types of content perform best on the Community Tab?

Home Videos Shorts Live Playlists **Community**

The Community Tab is a vibrant digital space where all manner of content can be incorporated.

However, some formats truly thrive in this environment.

- Polls are akin to planting the seeds of engagement - the more thoughtful the question, the richer the harvest of responses.
- Images and GIFs add splashes of colour, infusing your posts with personality and whimsy.
- But don't forget good old text updates to nurture that personal connection.

A balanced mix of these formats, and consistency, will ensure your community is always inviting!

Notes

QUESTION 5
How often should I post on the Community Tab?

Ah, this is a great question!

- Too little, and your community garden may wither from neglect.
- Too much, and you risk overfeeding to the point of disinterest.

As a general rule of thumb, **aim to add new posts every 2-3 days** - enough to keep things lively without overwhelming your viewers' feeds.

However, don't be afraid to adjust this cadence based on your audience's appetite.

Some followers may relish more frequent snapshots into your creative world, while others prefer a perfectly curated weekly update.

Listen to your viewers, experiment with different schedules, and you'll find the perfect rhythm for your blossoming community.

Most importantly - ASK your followers by posting as a question or a quiz.

Notes

QUESTION 6

Can I schedule Community Tab posts in advance?

But of course! Many Creators plan their strategies in advance, savvy YouTube Creators can plan their Community Tab posts well in advance too.

YouTube's clever little scheduling tool allows you to craft your updates, tend to the finer details, and set them to **go live on a future date and time** of your choosing.

This handy feature ensures your engagement efforts are never deterred by busy schedules or unexpected whimsies.

So go ahead:

- prepare those posts
- line them up
- set the publish dates
- and let your content publish automatically without having to fret over each trying to remember to post on a regular basis.

Notes

QUESTION 7
How can I use the Community Tab to promote my videos and grow my Channel?

While your videos are living on the YouTube Channel, the Community Tab offers the perfect place to nurture your viewers and followers!

- Share sneak peeks and teasers to build hype prior to new video launches.
- Pose thought-provoking polls or questions related to your content to spark discussions that lead viewers to explore your Channel further.
- Don't be afraid to promote your latest videos by sharing links, thumbnails and tantalising Descriptions.

When coupled with consistent, engaging community care, these practices will organically enrich your Channel's visibility and attract new members to engage and connect with your followers and fans.

Notes

QUESTION 8

Which analytics and metrics are available for Community Tab posts?

Luckily, YouTube provides a veritable wealth of insights about your Community Tab posts.

From views and engagement rates to traffic sources and audience territories, these metrics allow you to carefully analyse **which content resonates most with your viewers**.

Identify:

- which post formats perform the best
- which topics encourage discussion
- and which scheduling patterns yield the most engagement.

With each new data review, you'll be better equipped to thoughtfully cultivate a community experience tailored to your followers' unique tastes.

Notes

QUESTION 9

How can I encourage more engagement and interaction from my audience?

While clever, eye-catching posts attract your audience's gaze, it's **the element of interactivity that keeps them buzzing around your community**:

- ask questions that prompt deeper discussion
- share personal stories that forge heartfelt connections
- and never shy away from expressing your enthusiasm in the comments - a little digital sunshine can work wonders!

Initiate recurring themes like "Fan Art Fridays" or "Wild or Whacky Wednesdays" to encourage game-like engagement.

And for the boldest Creators, you could even open the floor to **fan-created post topics** every now and then!

An engaged community is a thriving one, so tend to those social connections with passion.

Notes

QUESTION 10

Can I collaborate with other creators or run contests/giveaways?

Oh, most definitely! In fact, **cross-pollination between communities** can truly unlock those big energies.

Collaborating with fellow Creators on **joint posts, live streams or challenge campaigns** allows you to experiment with new ground and introduce your great content to fresh eyes and personalities.

And what better way to celebrate growth milestones than by including epic giveaways and contests?

Not only do these events reward your existing followers, but their buzz spreads rapidly, attracting new visitors to join in on your Community Tab.

So start cultivating some cross-community magic!

Notes

QUESTION 11
Which unique or creative ideas can work in the Community Tab?

While the standard post formats offer boundless potential, the true magic happens when you **tap into your uninhibited imagination**!

Why not host "Riddle Me This" brainteasers where viewers must solve clever puzzles?

Or perhaps a "Wild Idea of the Week" spotlight on exceptional fan creations?

The sky's the limit for:

- unique series
- polls
- games and more that inject your distinct personality into the space.

Get those creative juices flowing and your community will be buzzing with activity!

Notes

QUESTION 12

How can I incorporate Community Tab content into my long-form videos?

Just as a talented designer incorporates existing patterns into new designs, you too can weave your community's voice throughout your videos!

- Highlight notable comments or fan art from the Tab during intro segments.
- Reference poll results or stitch together GIF reactions mid-video for an ultra-engaging experience.
- You could even solicit video ideas prompted by Tab responses and give Creators shout-outs!

An immersive cross-pollination between videos and community is sure to delight your loyal followers.

Notes

QUESTION 13
Are there any interesting case studies of creators successfully using this?

Home Videos Shorts Live Playlists **Community**

Yes there most certainly are interesting Creators who have cultivated wonderfully engaging communities through inspired use of the Community Tab!

- Channels like **The Game Theorists (Matthew Patrick)** foster epic narrative discussions by seeding lore-steeped polls and teasers.
- https://www.youtube.com/@GameTheory/community
- Andrew Rea of **Binging with Babish** whets his culinary tribe's appetite with behind-the-scenes previews.
- https://www.youtube.com/@babishculinaryuniverse/community
- **MrBeast** frequently uses polls and questions on the Community Tab to get feedback from his massive audience on video ideas, charity initiatives, and more. This direct engagement helps shape his popular content.
- https://www.youtube.com/@MrBeast/community

- In addition to posting updates, **Markiplier** shares exclusive behind-the-scenes images/GIFs from his videos and skits, building excitement with his community.
- https://www.youtube.com/@markiplier/community
- **Jackie Aina** the beauty guru uses the Tab for candid chats, asking her audience for advice on upcoming videos/products and making her fans feel involved.
- https://www.youtube.com/@jackieaina/community
- **The LaBrant Family** vlogging channel posts fun sneak peeks, teasers and conversational prompts to keep their audience engaged between videos.
- https://www.youtube.com/@ColeAndSav/community
- **Mark Rober** of the **squirrel videos fame** is one of my all-time favourite YouTubers :-) He's not posting too often on his Community Tab, but his posts are thought-provoking and interesting.
- https://www.youtube.com/@MarkRober/community

These are just a few of the amazing YouTube channels who are using the Community Tab to great advantage.

By cross-pollinating unique creativity with consistent nurturing, the seemingly ordinary Tab becomes a vibrant community!

Notes

Notes

QUESTION 14

What are some dos and don'ts for Community Tab etiquette and post formats?

When communing with your Community residents, be sure to mind your manners!

- Do reply to comments, pose thoughtful follow-ups and express gratitude for the engagement activities.
- Don't, however, over-post by replying to every single "reaction" - that leads to potential spam choking the conversation.
- Do evaluate post formats through an accessibility lens by providing descriptive text for images.
- But don't overdo it by bombarding the viewers with too many posts per day.

As with any community, mindful upkeep and natural moderation will bring the most vibrant results!

Notes

QUESTION 15

Can I respond to Community Tab comments/posts on a mobile device?

But of course! YouTube's mobile app experience allows you to connect to your community at any time, from anywhere, on mobile devices.

Not only can you view and engage with existing posts, but the app allows effortless creation of new updates as whimsical inspiration strikes!

Whether capturing enchanting selfies or sharing fun and interesting status updates, your mobile device can definitely be part of your routine.

You will definitely appreciate the added portability.

Notes

THAT'S THE END of the 15 questions on the Community Tab. Did you learn something new?

Home Videos Shorts Live Playlists **Community**

Now you've checked out those questions, I have a few more tidbits for you.

If you've been wondering about using the community tab, and how to tell people about it, on the next page I'm including:

- **8 Simple Ways to Spread The Word About The Community Tab**.

After that, you'll find:

- **15 Ideas for Effective Thumbnails, Titles, and Engaging Content for Community Tab Videos**.

8 Simple Ways To Spread the Word About the Community Tab

1. The Trailer Tease
Create a tantalising video trailer hyping up the behind-the-scenes magic happening in your Community Tab. Sprinkle in snippets and sneak peeks to whet your viewers' appetites!

2. In-Video Signposts
Why not beckon your tribe of followers to the Community Tab directly from your videos? Add creative call-to-action (CTA) elements like fun lower thirds or zany mid-roll alerts pointing them to the community engagement.

3. Social Media Murmurings
Let the whisperings of your Community Tab adventures drift through social channels. Share tantalising glimpses on Twitter, Instagram, Pinterest, website, blog and beyond to pique curiosity.

4. The Loyalty Reward
Shower your most devoted fans and followers with an exclusive Community Tab perk or inside scoop as a token of appreciation for their viewership!

5. Collaborator Cross-Pollination
When joining forces with fellow YouTube Creators, have them give a shout-out to your welcoming Community Tab during videos or live streams.

Notes

8 Simple Ways To Spread the Word About the Community Tab, Cont:

6. Real-World Breadcrumb Trail
For an extra whimsical touch, leave a real-world trail of clues and curiosities pointing to your online Community Tab! Imaginative guerilla marketing can add an air of mystique.

7. Let Your Branding Bloom
Infuse your channel branding and artwork with subtle nods to your Community Tab. An elegant word, symbol or mascot could become a playful breadcrumb-enticing exploration.

8. News and Clues
If you have a newsletter or list of any kind, let your followers know they can connect on the YouTube Community Tab, rather than having to make a comment on a video already on the Channel. This is a special way to build loyalty and trust with followers when they know they can reach out and connect easily via the Community Tab.

The key with using the Community Tab is to get creative and have fun with it!

Making your community space feel like a vibrant, exclusive experience will organically attract your lovely tribe of engaged viewers.

Notes

15 Ideas for Effective Thumbnails, Titles, and Engaging Content for Community Tab Videos

5 Most Effective Thumbnails:

1. Close-up of a smiling face - creates a personal connection
2. You holding or pointing to relevant props - draws curiosity
3. Bold contrasting fonts/colours - grabs attention
4. Image representing video topic - hints at content
5. Puzzle or question imagery - makes viewers want to watch to find answers.

Most Effective Titles:

1. Ask a compelling question - makes viewers want to get the answer
2. Use viewer's name - creates a personal feeling of being addressed
3. Include numbers or lists - appeals to viewer's desire for specifics
4. Use "You won't believe ..." - builds intrigue and interest
5. Starting with "The secret to ..." or "How to ..." gives value and helps solve problems.

Notes

15 Ideas for Effective Thumbnails, Titles, and Engaging Content for Community Tab Videos, Cont:

Most Engaging & Interesting Content:

1. Share personal stories and experiences - this allows viewers to relate to you
2. Give step-by-step instructions or tips - this provides value for viewers
3. Answer common questions from viewers - this addresses what they want to know
4. Share behind-the-scenes looks - this gives a unique insider's view
5. Reveal your mistakes/lessons - this shows your vulnerable, real side.

I hope these ideas and tips have your creative juices kicking in and getting excited about using some of them on your own YouTube Channel!

Have fun!

Notes

Do you have a question for Teena?

Now that you've read through this booklet, have you made notes and have questions for me? Fantastic!

Pop over to this link:

click here to send me an email >>

What Teena's Clients & Students are saying

Teena, I cannot wait for you to share your expertise on the YouTube Community Tab with more Creators around the world! This book is sure to be an invaluable resource, nurturing vibrant online communities through your imaginative ideas and actionable advice.
Claude A. AUS

I literally can't wait to put these tips and shortcuts into practice! This will be a total game changer for me! Thanks so much for sharing Teena! :)
Marnie M, FR

Thank you for the inspiration to help my YouTube Channel! TEENA YOU ROCK!!
Mark V, USA

Teena! These questions and answers made it so easy to understand. Just in case someone didn't tell you today, or lately, you rock!
Frederic, FR

What Teena's Clients are saying

Thank you, thank you, thank you for your incredible ideas about the Community Tab for my Channel! The time you've spent writing this booklet to explain it all to help folks like me is amazing. I can't express how grateful I am to have your support via this booklet.
Shereen B. USA

Excellent explanations! I was sharing these concepts with my virtual assistant, and to make extra certain that I was teaching her correctly, I bought her a copy of your booklet as well. Thank you!

Michelle N. UK

Just grabbed a copy of the book and looked it over and I am gobsmacked. This is so good and packed with power.
Eloise, CAN

Wow! These tips are absolutely brilliant, Teena! Literally can't wait to put them into practice! Definitely will start doing this every time I post a new video!
Meredith R, GR

Teena's Other Books and Training Courses

Book: 50 Marketing Secrets of Successful Women

- https://amzn.to/48XkCj0

Book: 50 Quick & Easy Website Tips

- https://amzn.to/49T49h0

Freebies, PDFs, Cheat Sheets, Checklists

- https://teenahughesonline.gumroad.com

Training Courses

- https://teenahughesonline.com/training-courses-by-teena-hughes

Thank you so much for buying this book, I do hope it helps you build your YouTube Channel and find your followers 🖤

THE END

❤

Made in United States
Troutdale, OR
11/03/2026

56844671R00037